Our Natural Resources

Written by Debbie O'Brien

Table of Contents

Natural Resources

Natural resources are found in nature. Trees, water, and rocks are **natural resources.**

Thanks to natural resources, you have water to drink, books to read, and lights to turn on. Let's find out more about how we use them.

This old mill was once powered by water.

Trees

Did you know that anything made out of paper comes from the wood of a tree?

Wood is used to make everything from pencils to furniture to houses.

logging truck

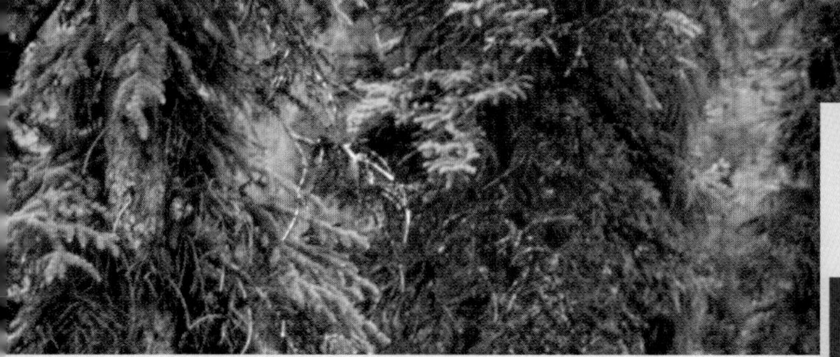

The frame of this house is made of wood.

Trees are also important because they help to make and clean the air we breathe.

Two healthy trees can make a year's worth of oxygen for one person.

Water

All living things need water. It is one of very few things that people cannot live without. We need fresh water to drink and to grow the foods we eat.

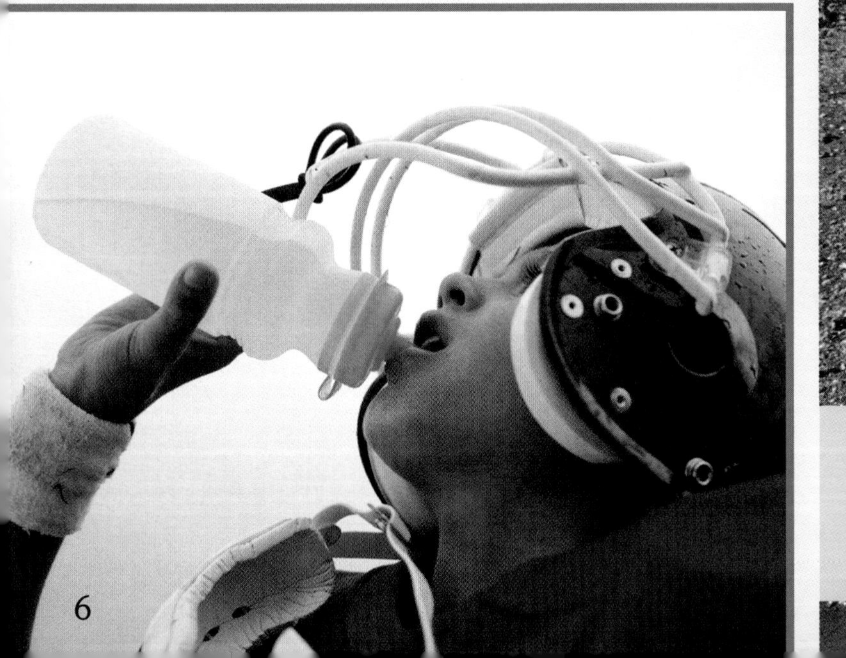

In dry areas, farmers use irrigation to water their crops.

bears fishing for salmon

Most of Earth's water is salty ocean water. But we need fresh water. Most fresh water comes from rivers, lakes, and ponds.

Water helps people in many ways. Some of the electricity that we use is made by water power. It is called **hydroelectricity** and is made at large dams.

Think about all the ways you use water.

city lights

Falling water passes through a dam and turns the motors that make electricity.

Reservoirs created by dams also provide recreation areas.

Rocks

Rocks may not sound important—but they are. That's because rocks are made up of **minerals.** Minerals are used to make everything from mirrors to toothpaste to tools.

This underground gold mine is 1,000 feet deep.

stone lined with quartz

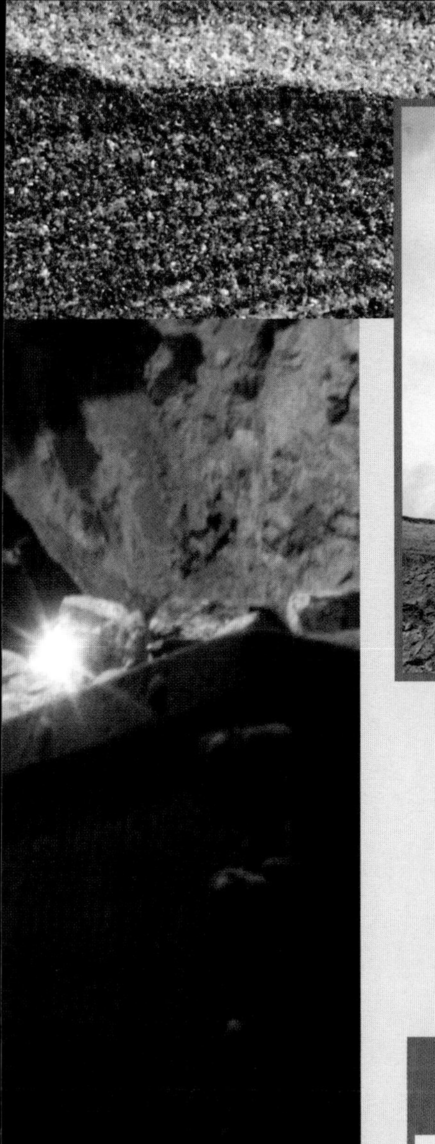

open pit mine

Miners often must dig deep into the earth to get rocks and minerals.

Mineral ➡	How It's Used
salt	spice, melt ice
quartz	watches, jewelry, television transmitters
iron	steel for cars, nails, appliances
limestone	toothpaste, cement

Fuels

Oil, natural gas, and coal are natural resources called **fuels.** We use these fuels to run our cars, to heat our homes in winter, and to cook our food.

gas stove

This oil drilling platform is in the ocean.

The gasoline in these cars is made from oil.

These fuels took millions of years to form, and they are buried deep below Earth's surface. It takes hard work to get them out of the ground.

Taking Care of Resources

Some natural resources are **renewable.** They can be replaced over time. Trees and plants are renewable resources.

tree seedlings

Minerals, oil, and natural gas are **nonrenewable** resources. Once they are used up, they are gone forever. To save these resources, scientists look for other energy sources, such as wind and sun.

Solar panels on a roof generate electricity.

wind farms

It's our job to take care of our natural resources. What can you do to help?